WOLF

COLORING BOOKS FOR ADULTS

STRESS RELIEVING PATTERNS

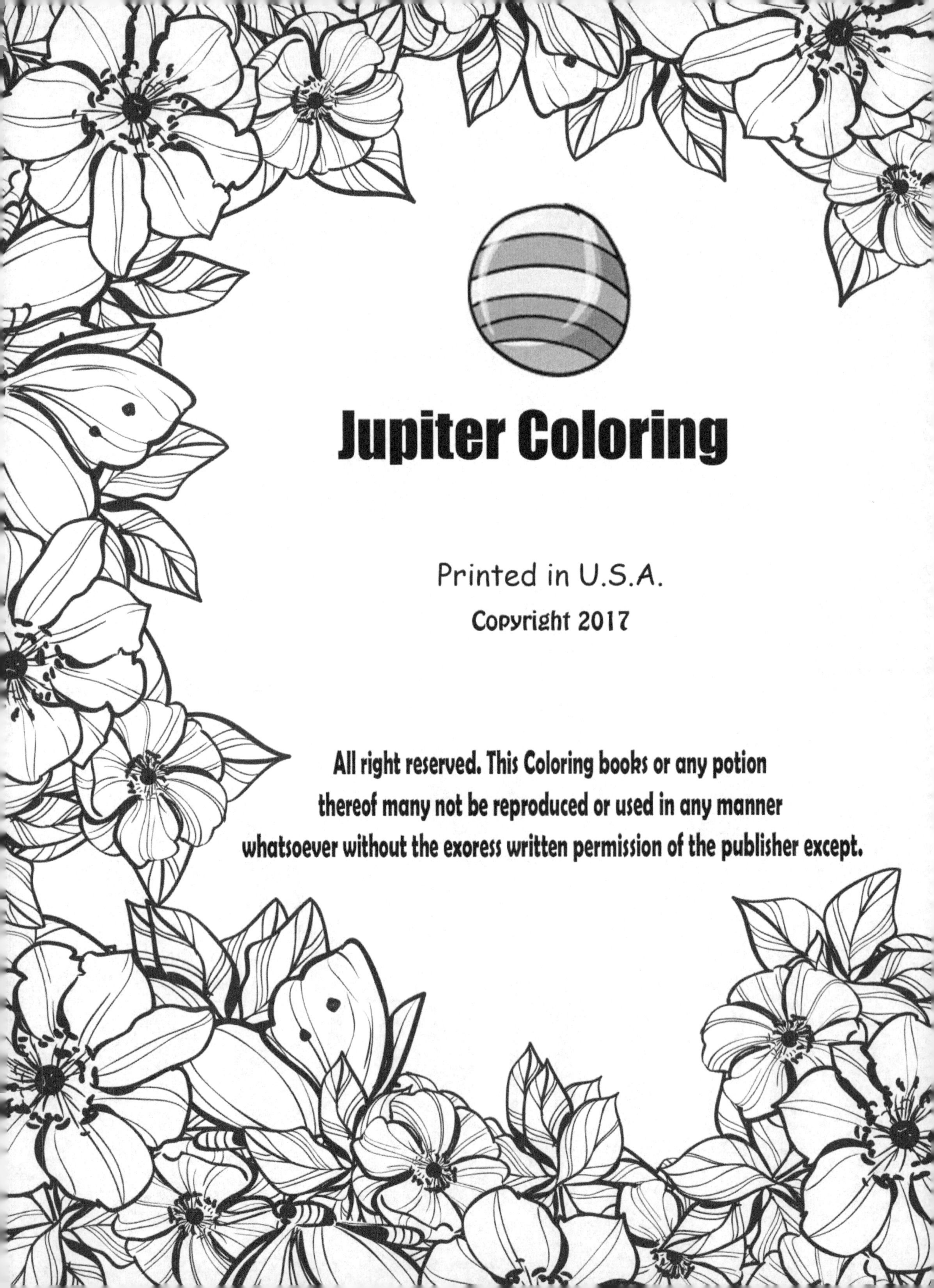

Jupiter Coloring

Printed in U.S.A.

This book belongs to:

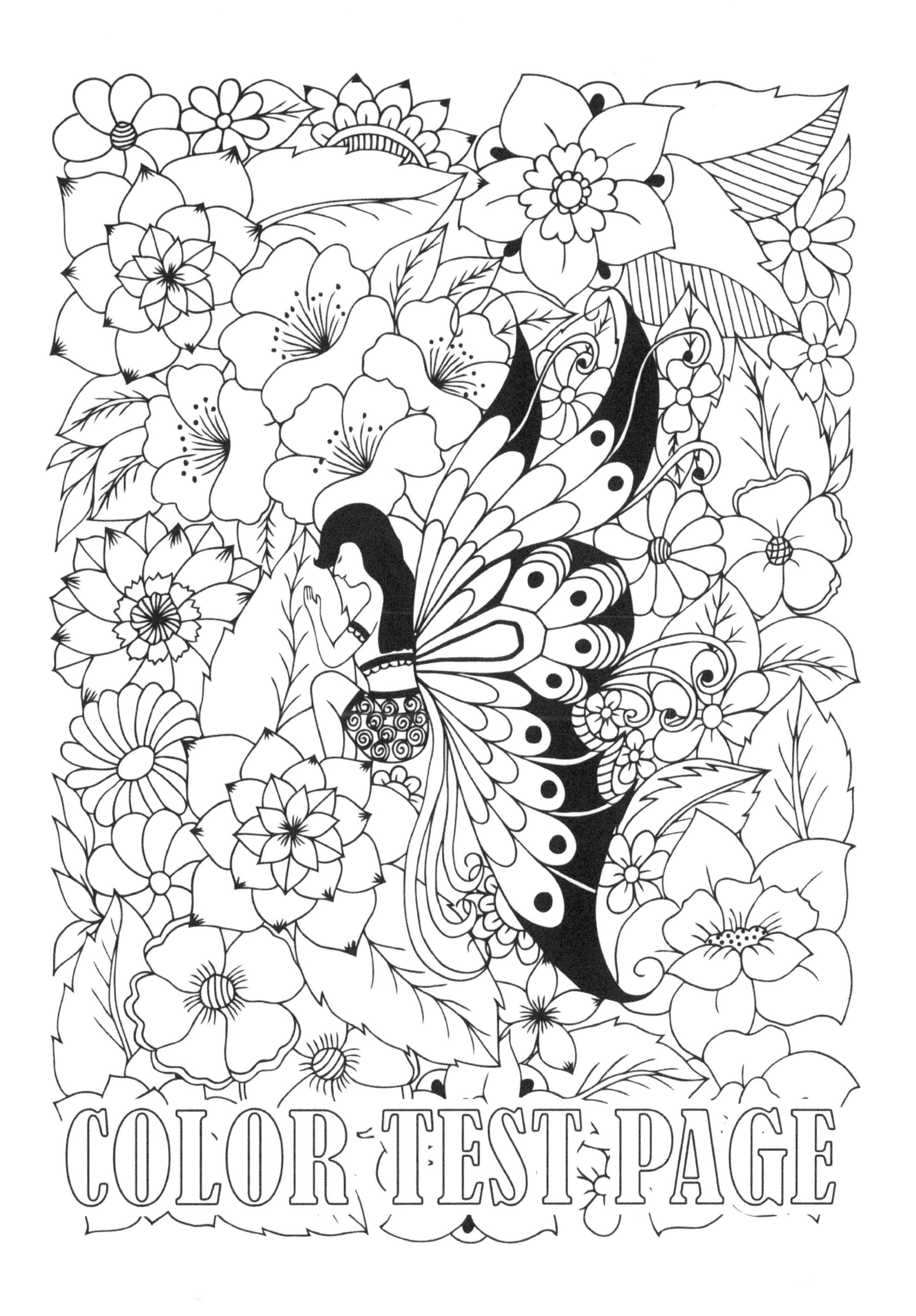
COLOR TEST PAGE

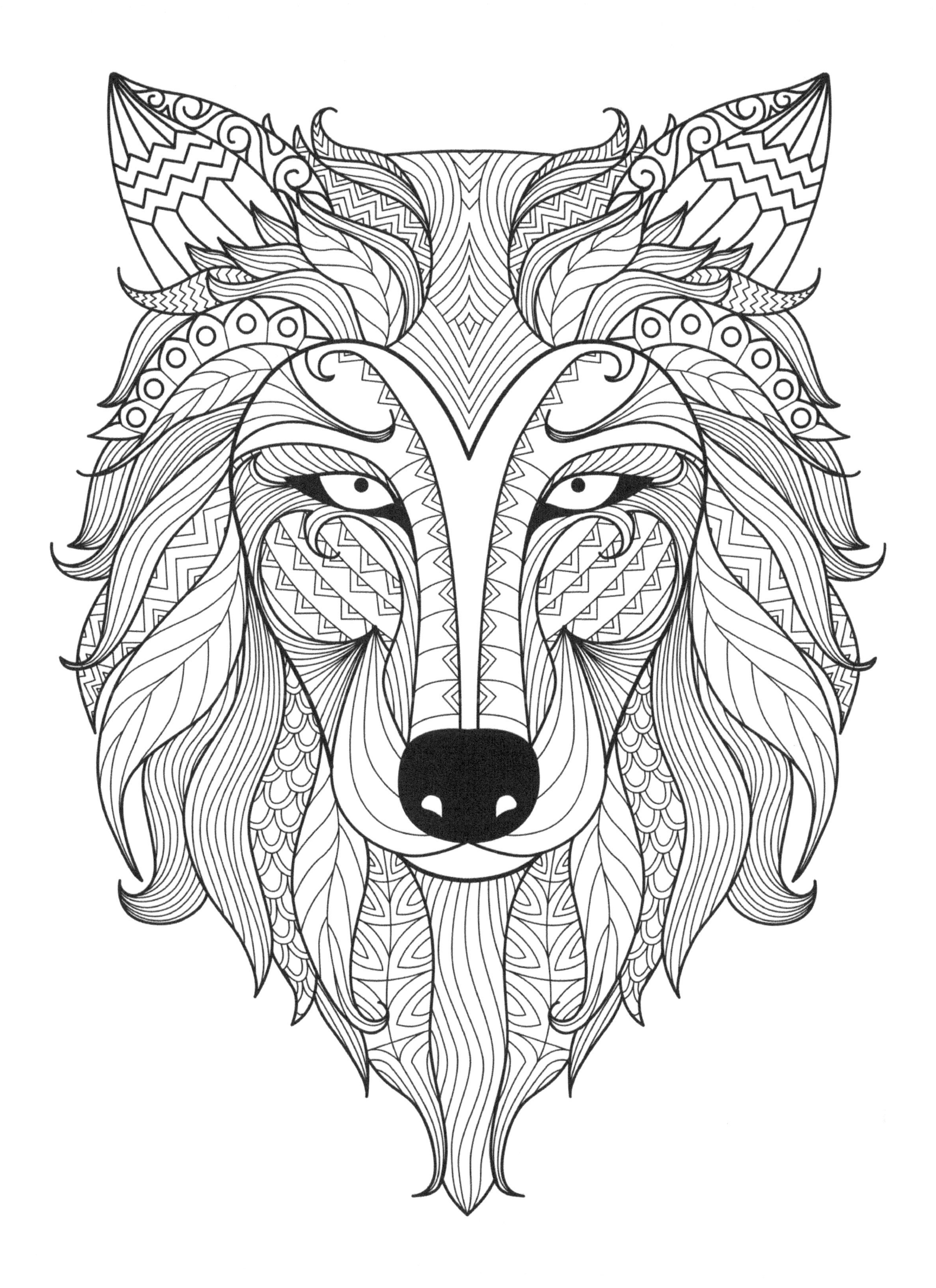

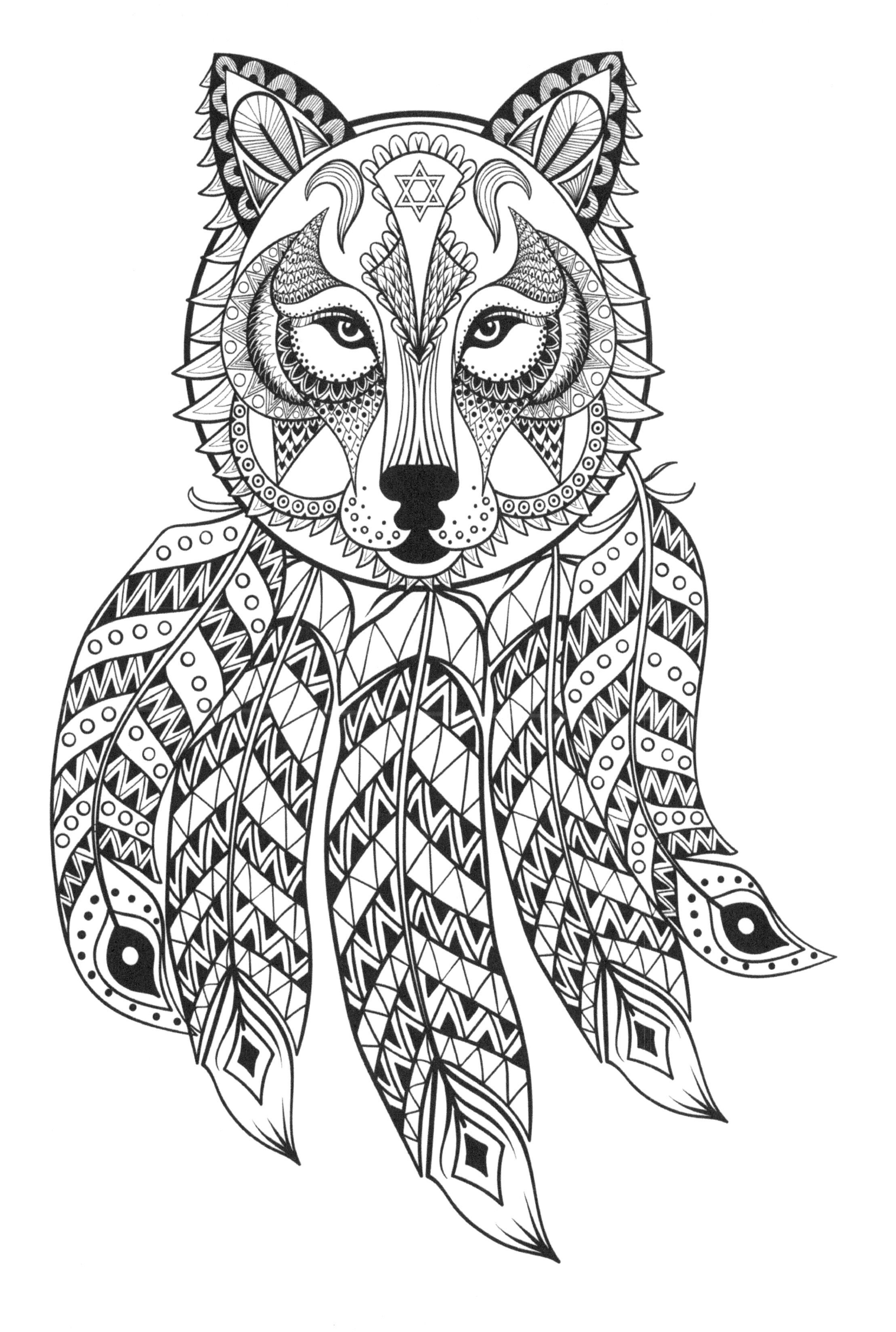

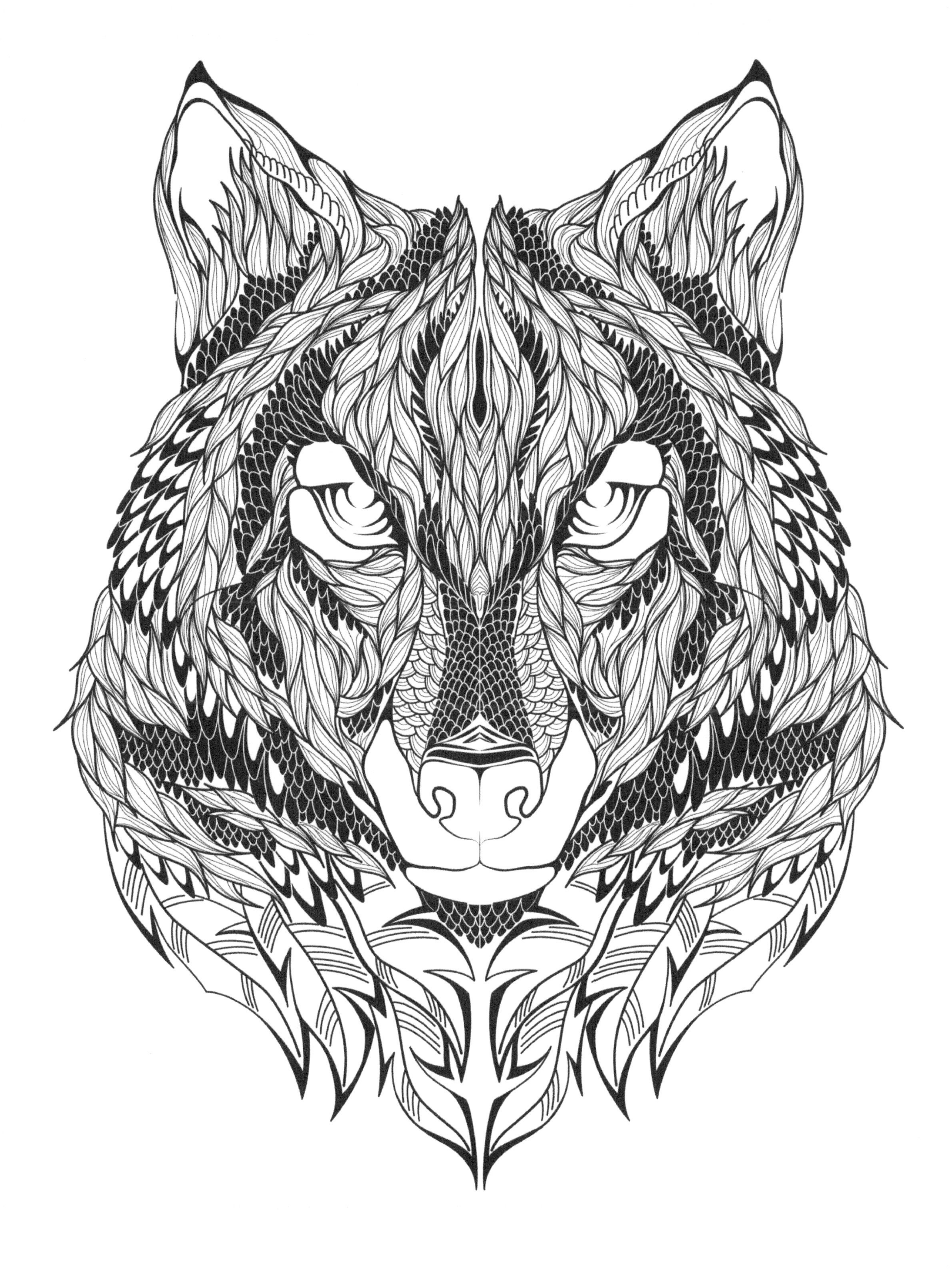

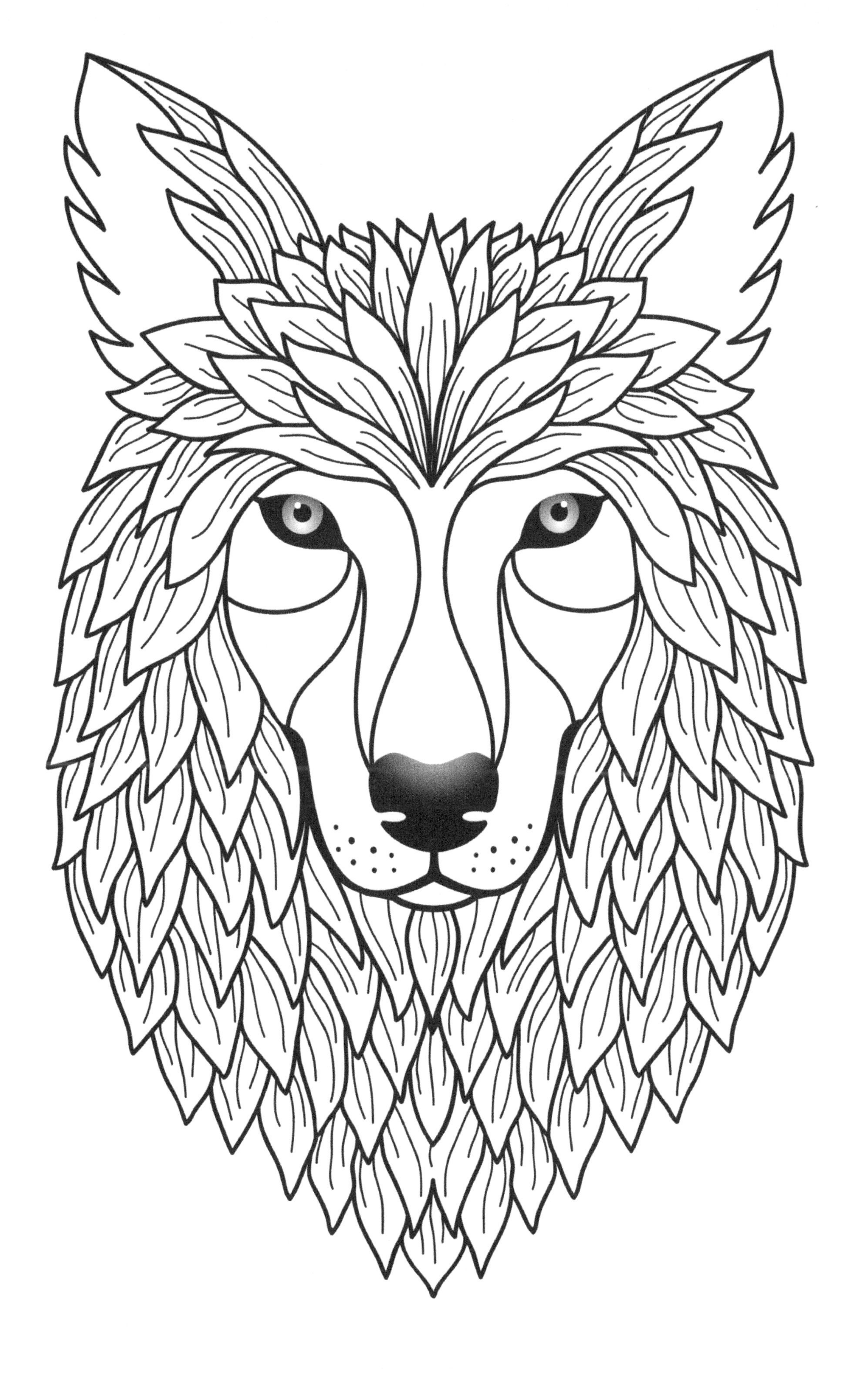

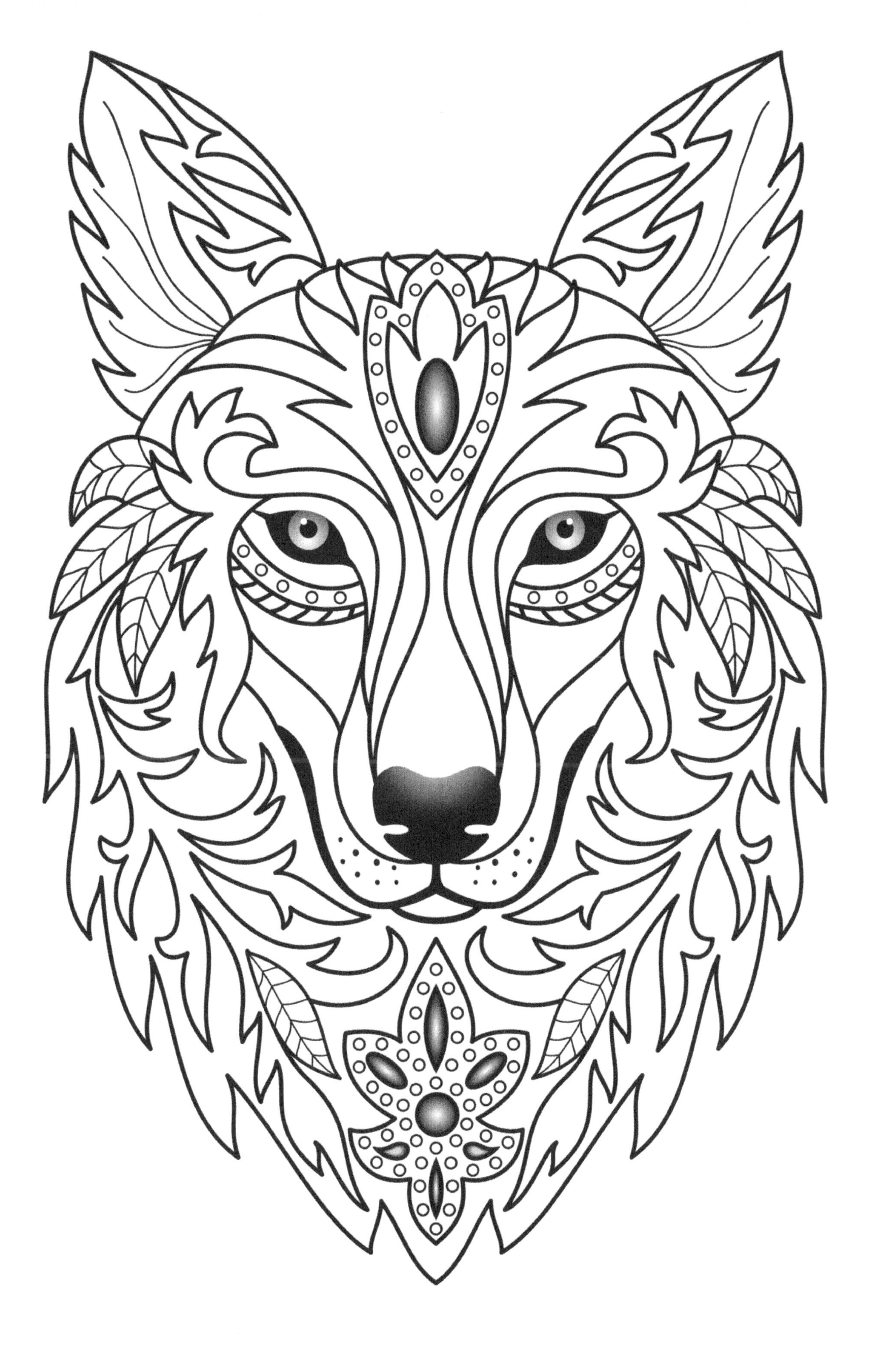

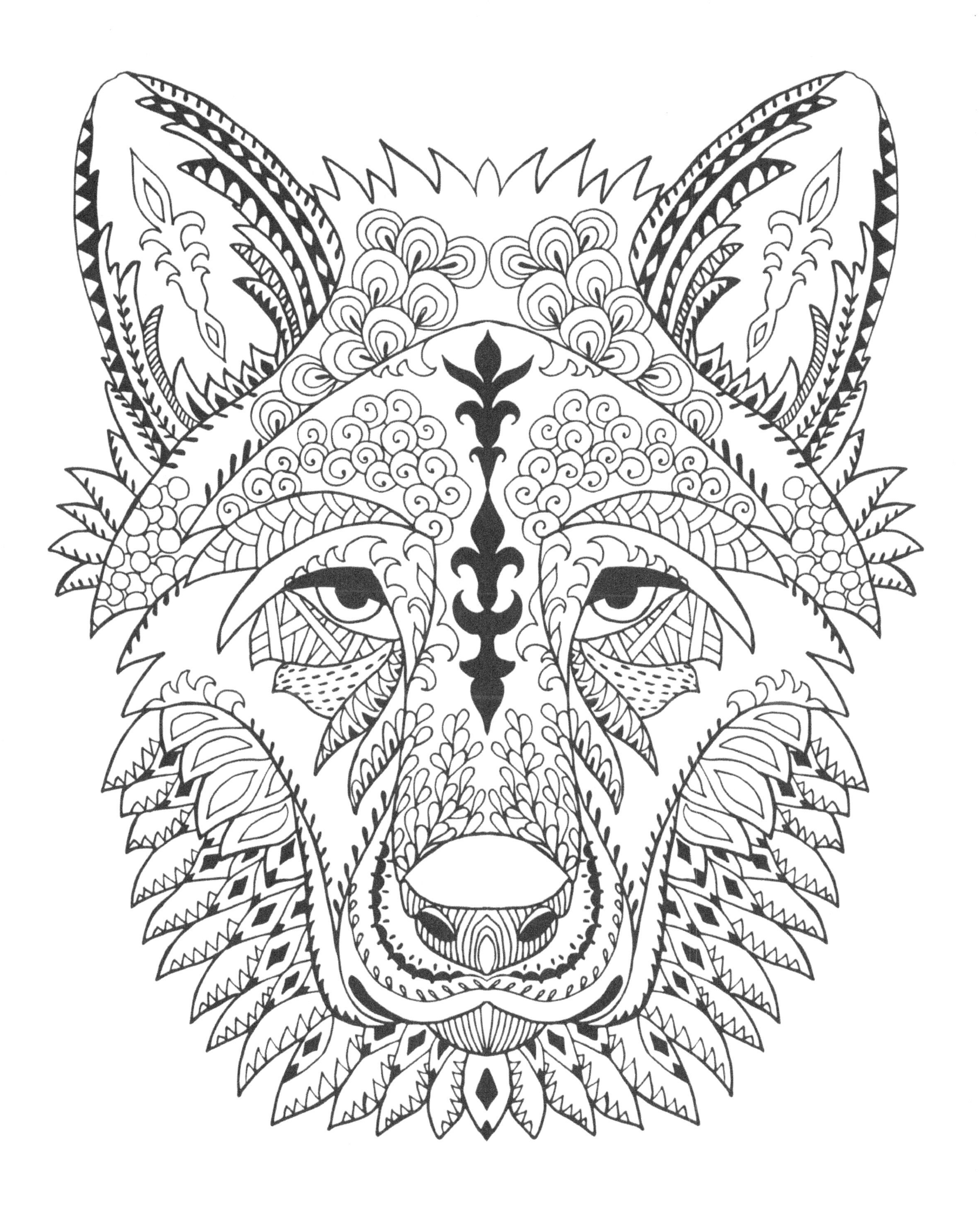

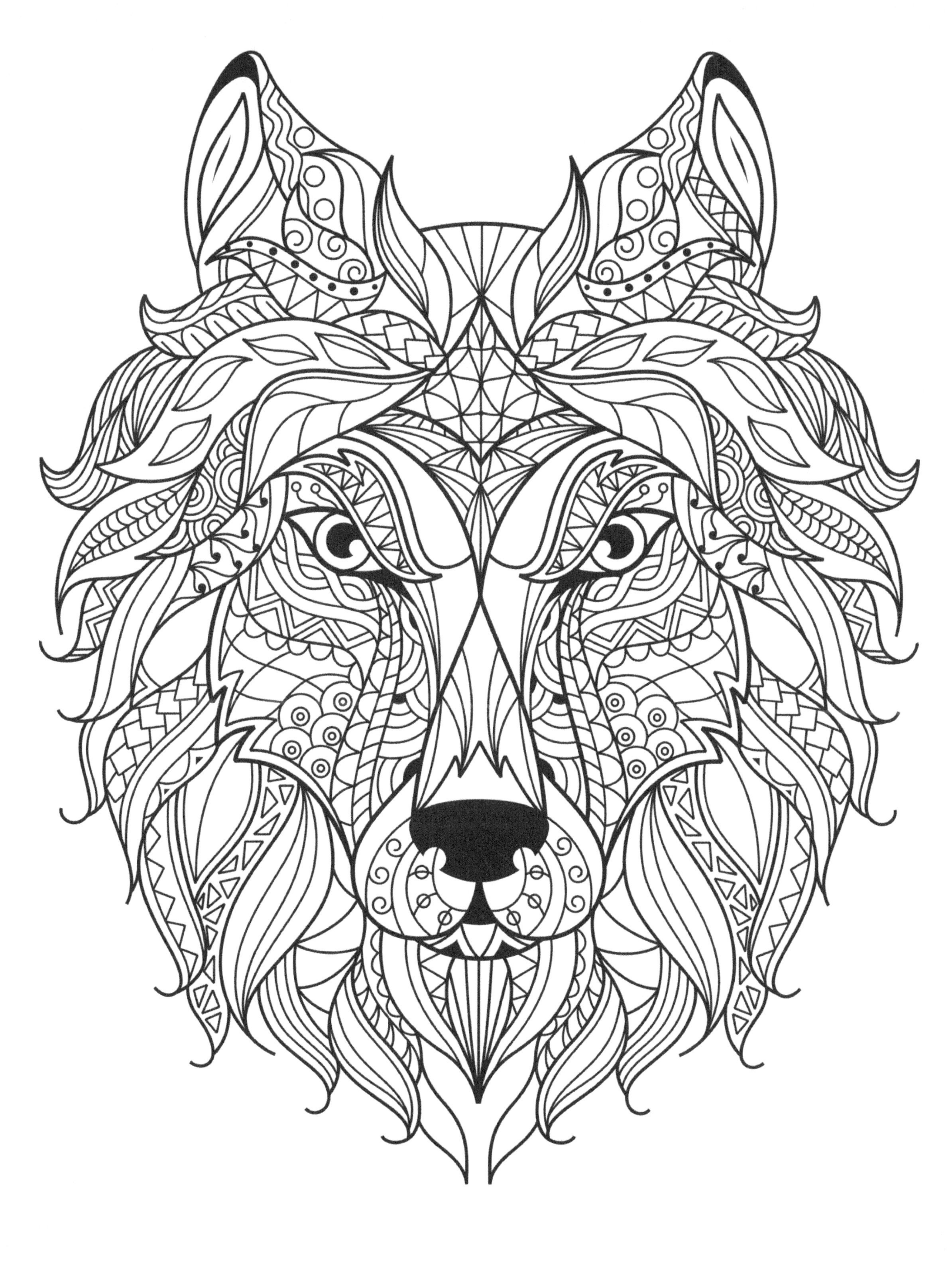

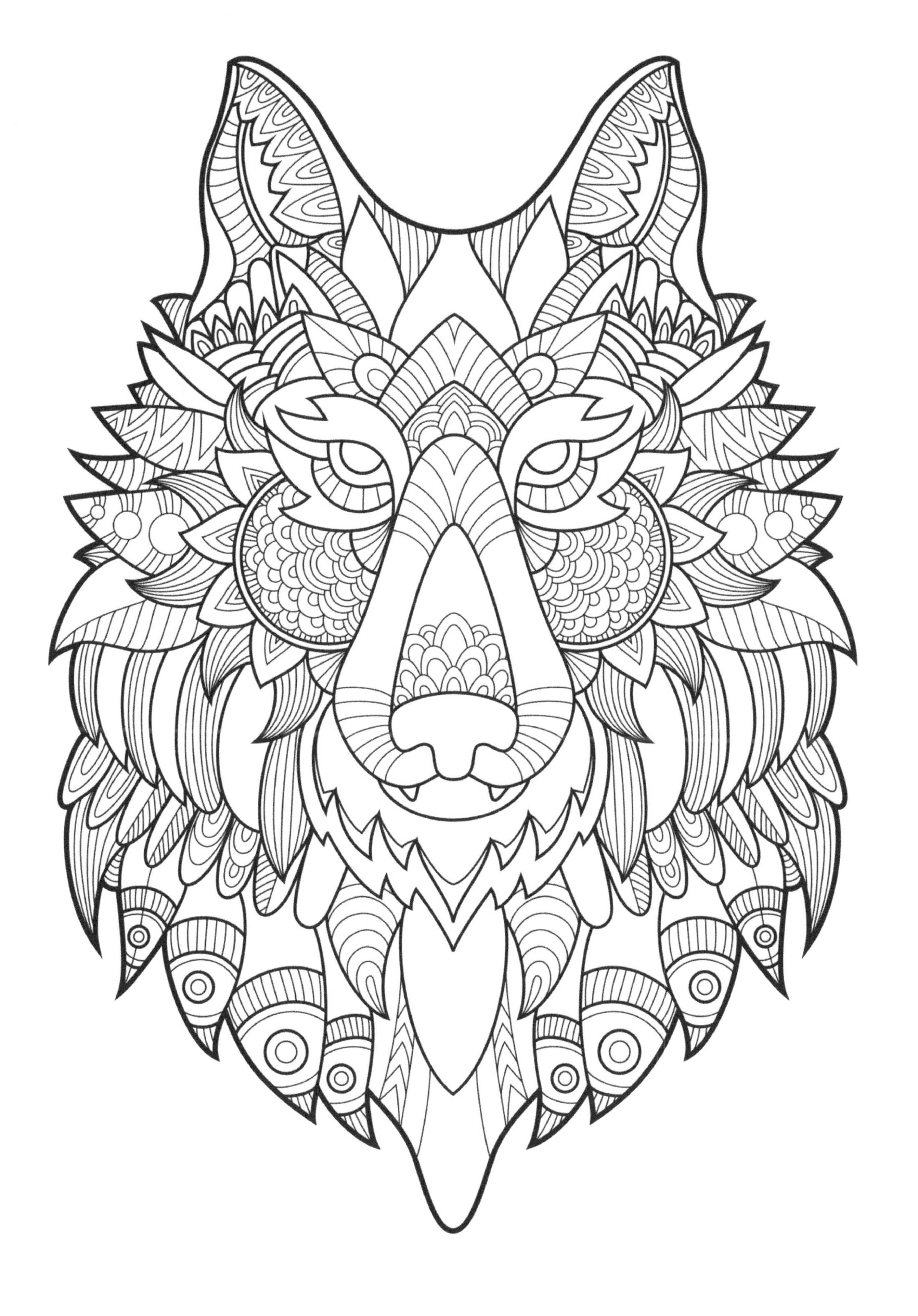

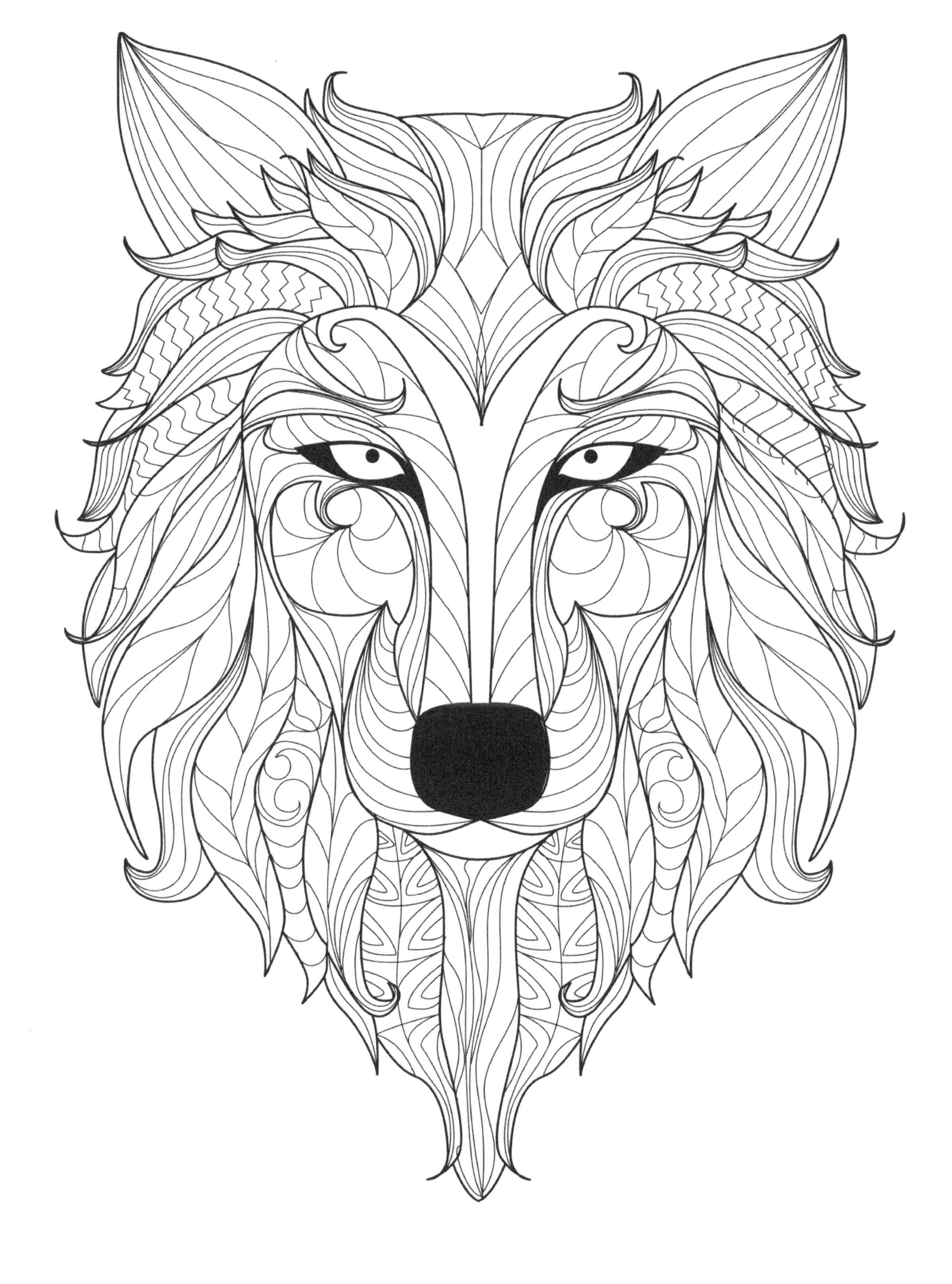

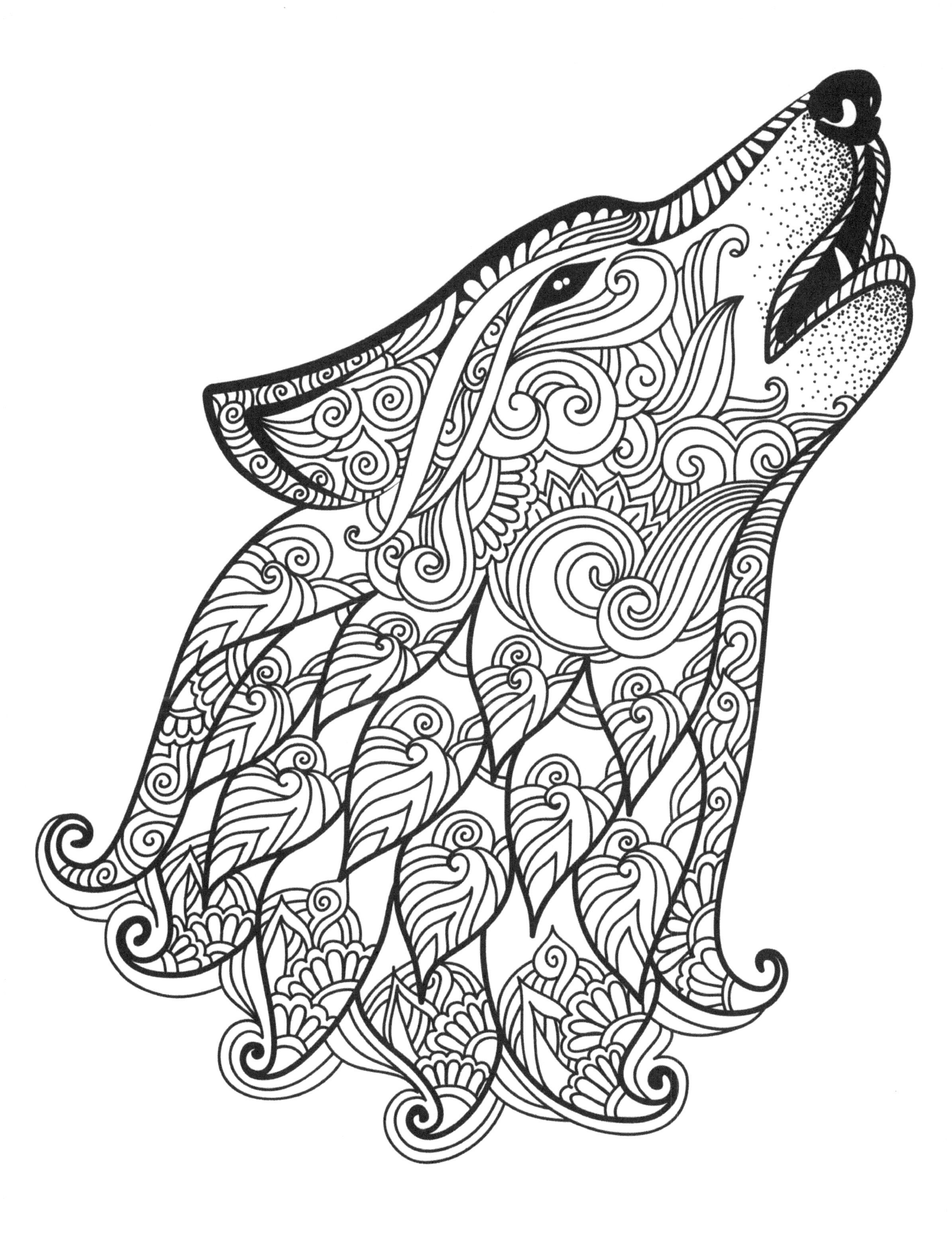

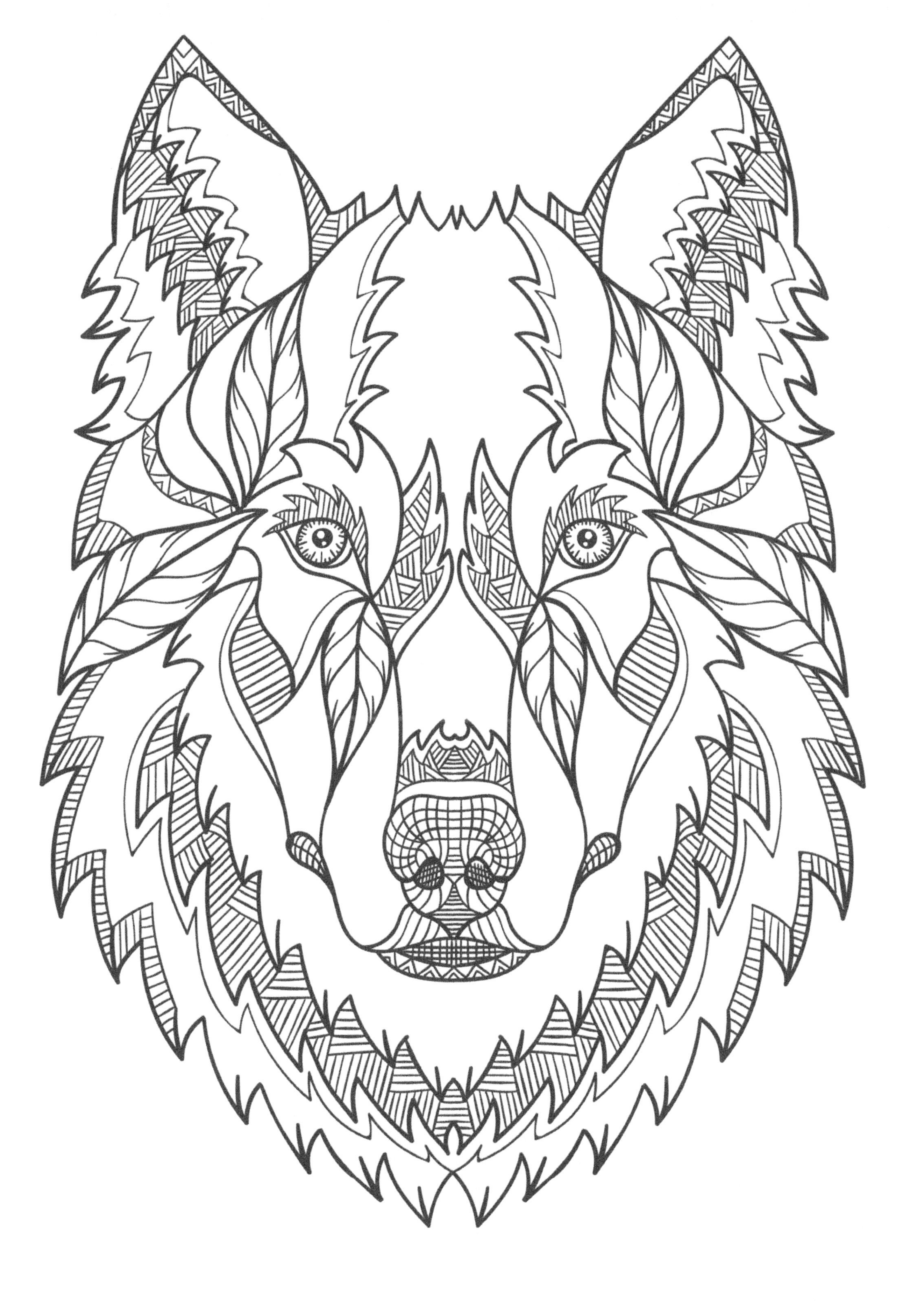

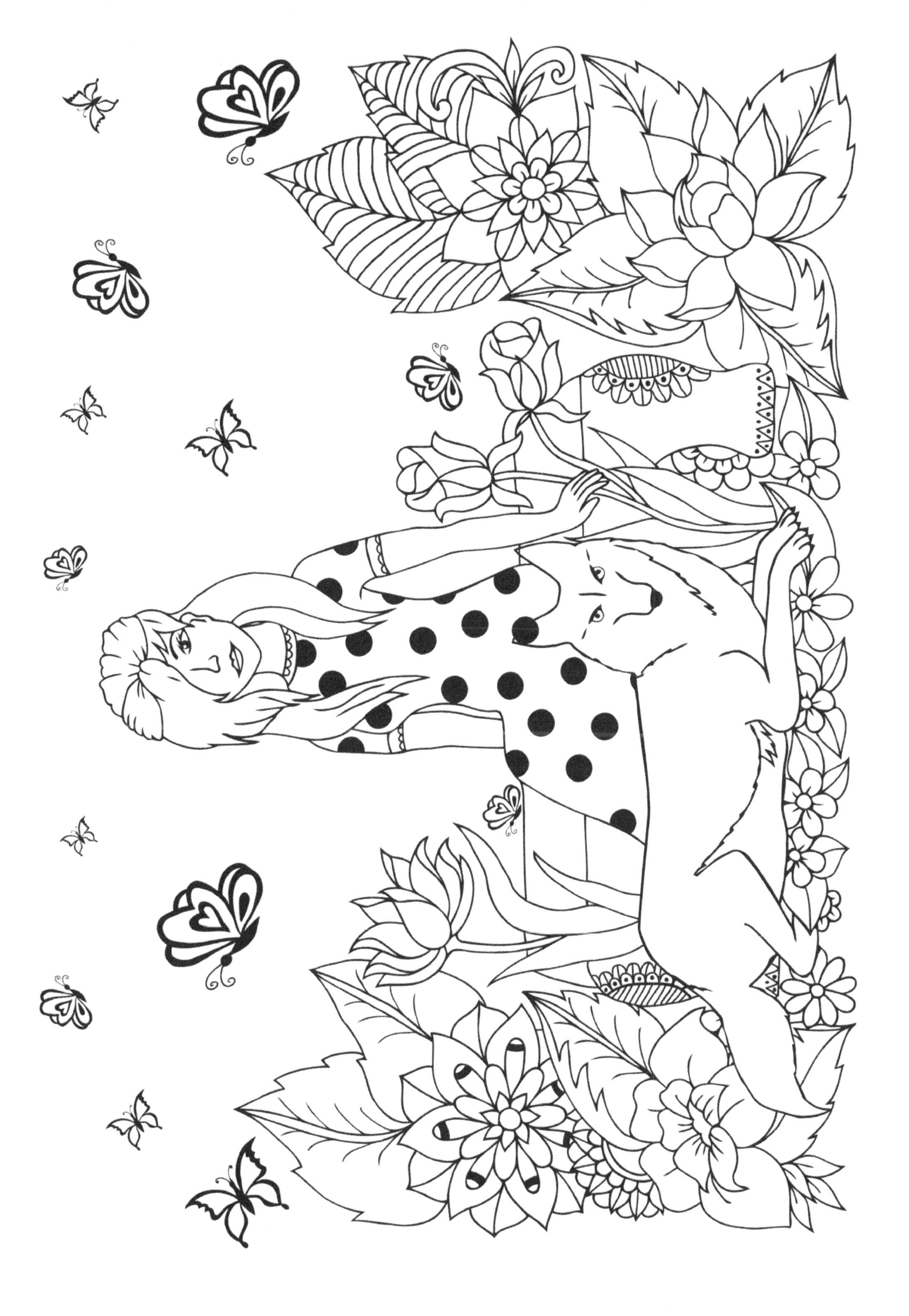

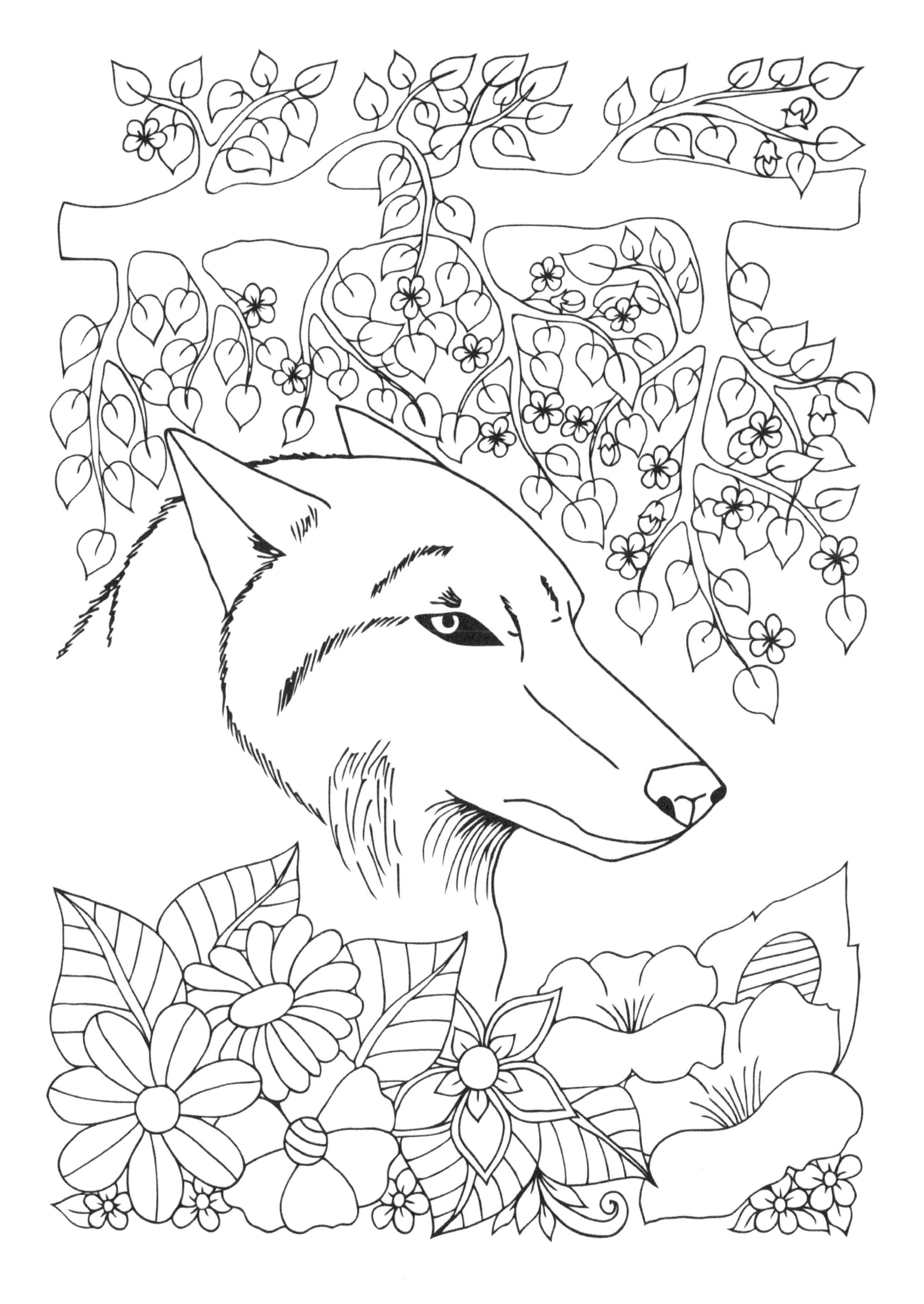

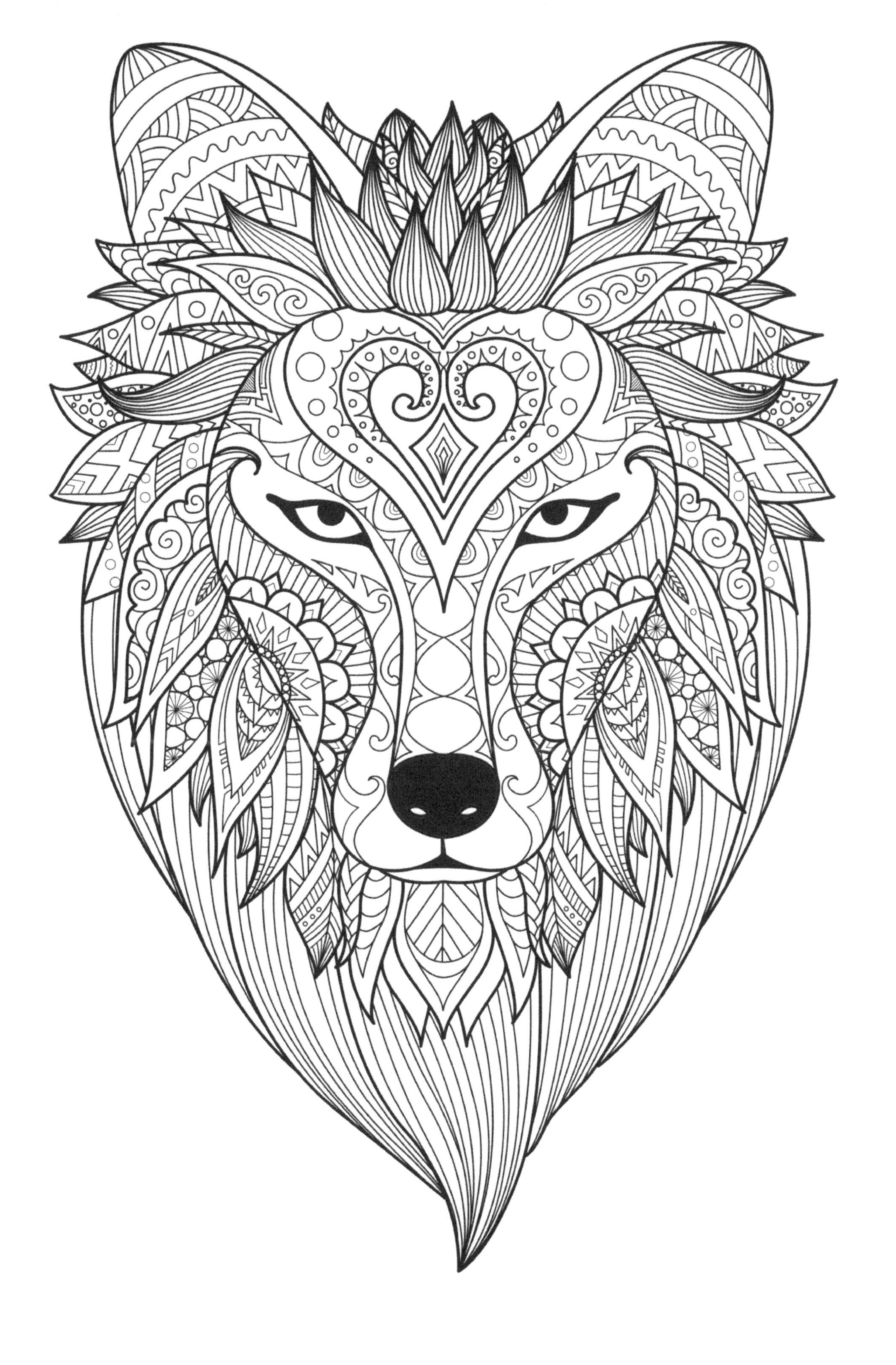

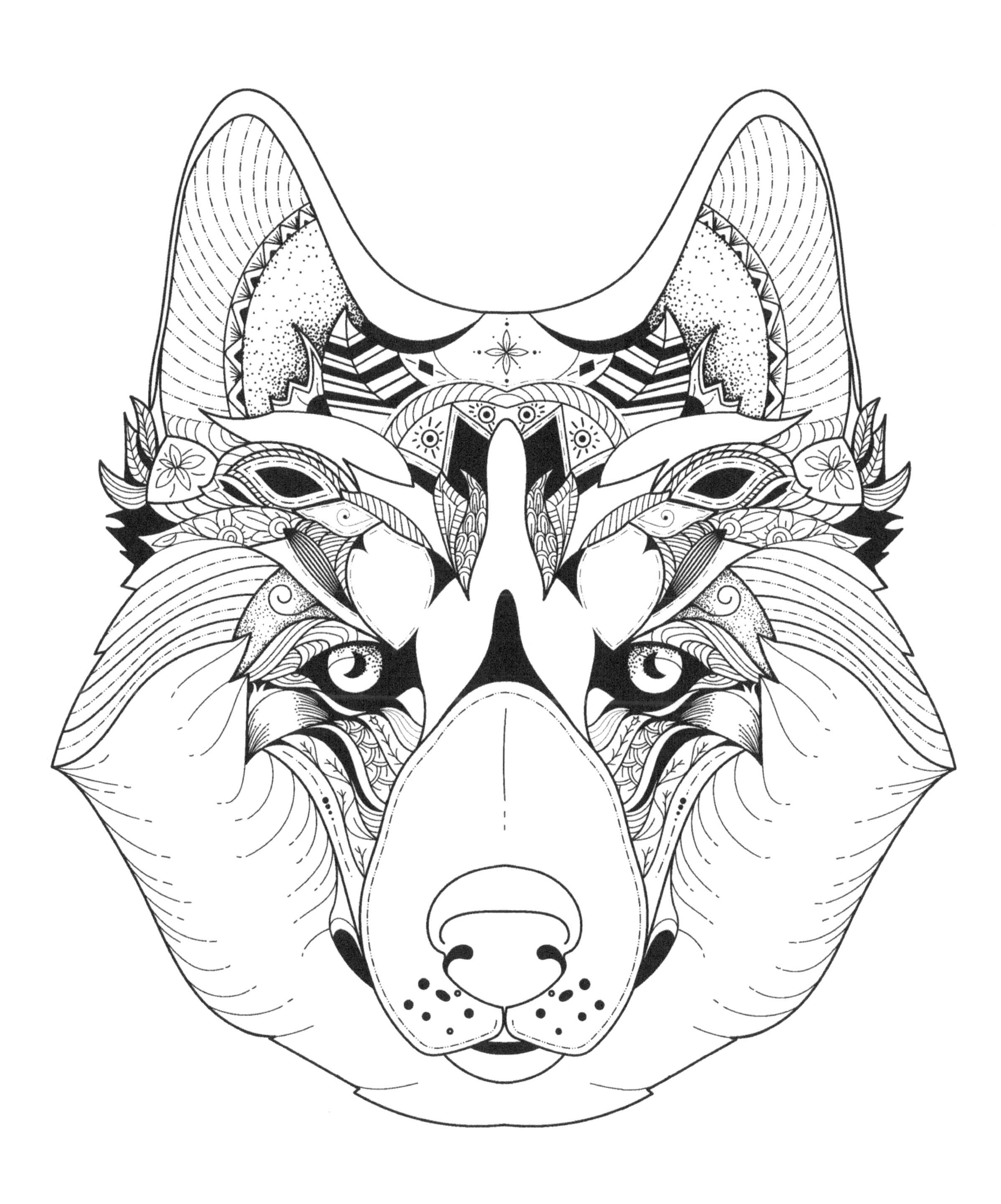

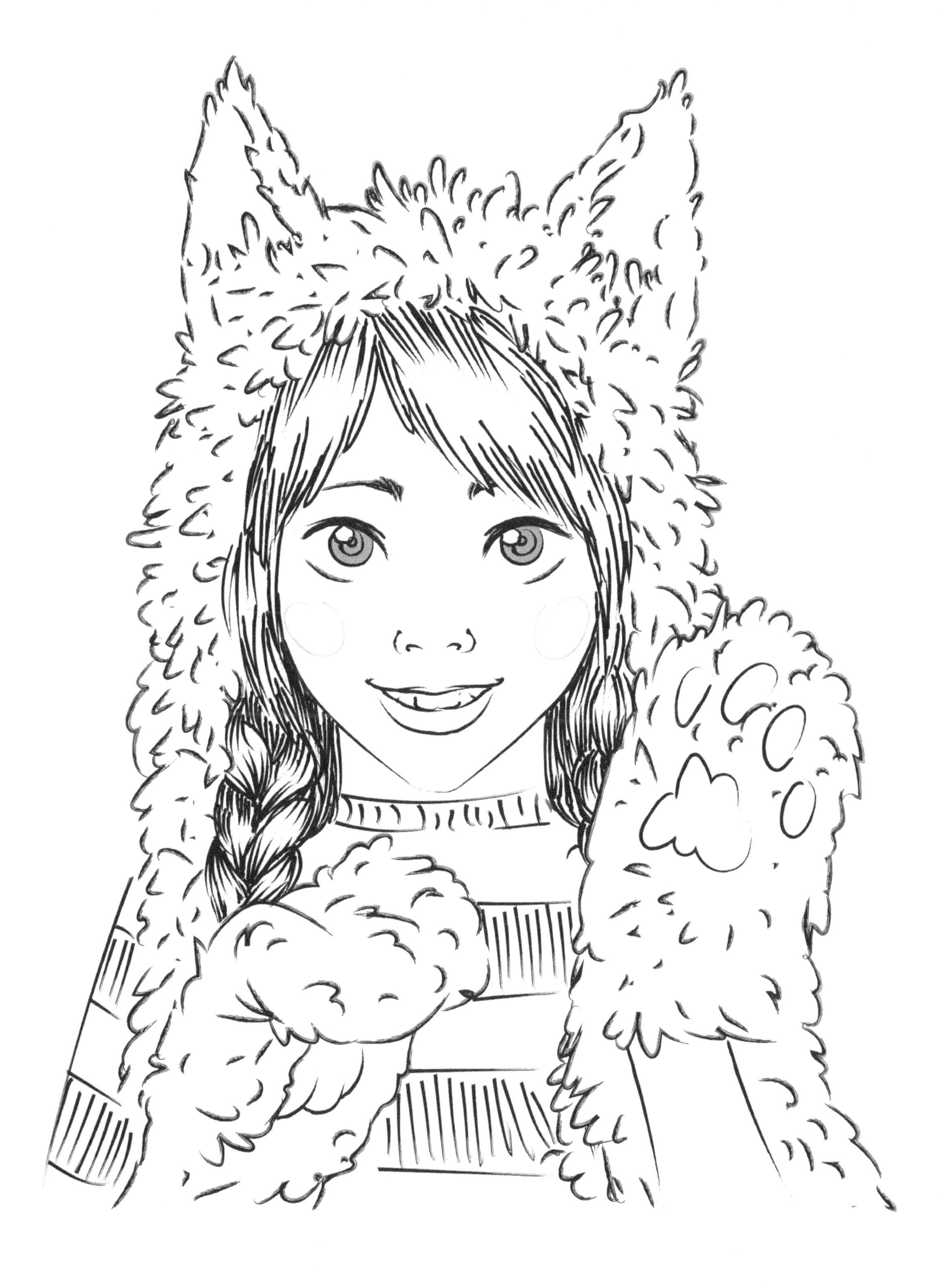

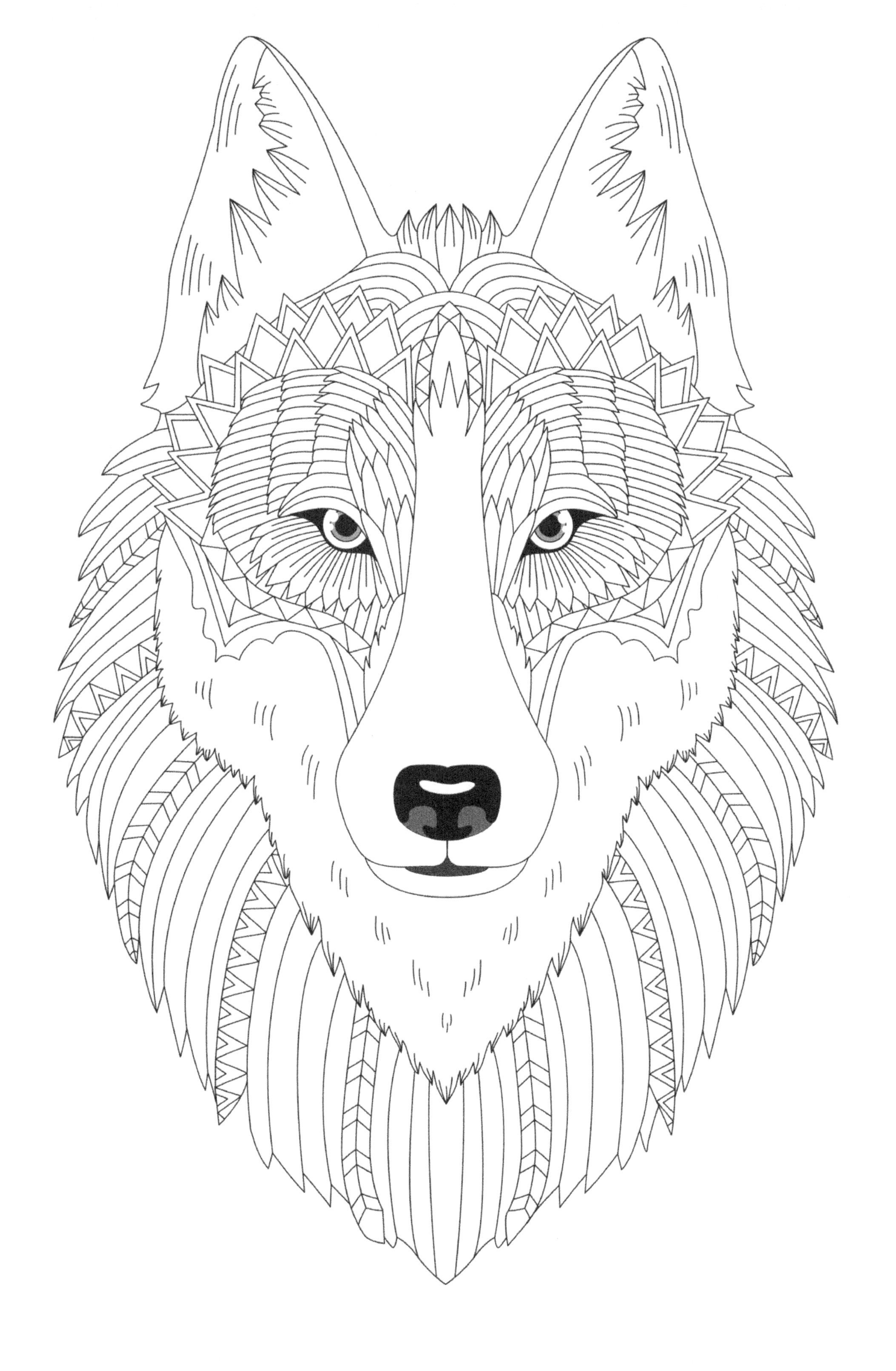

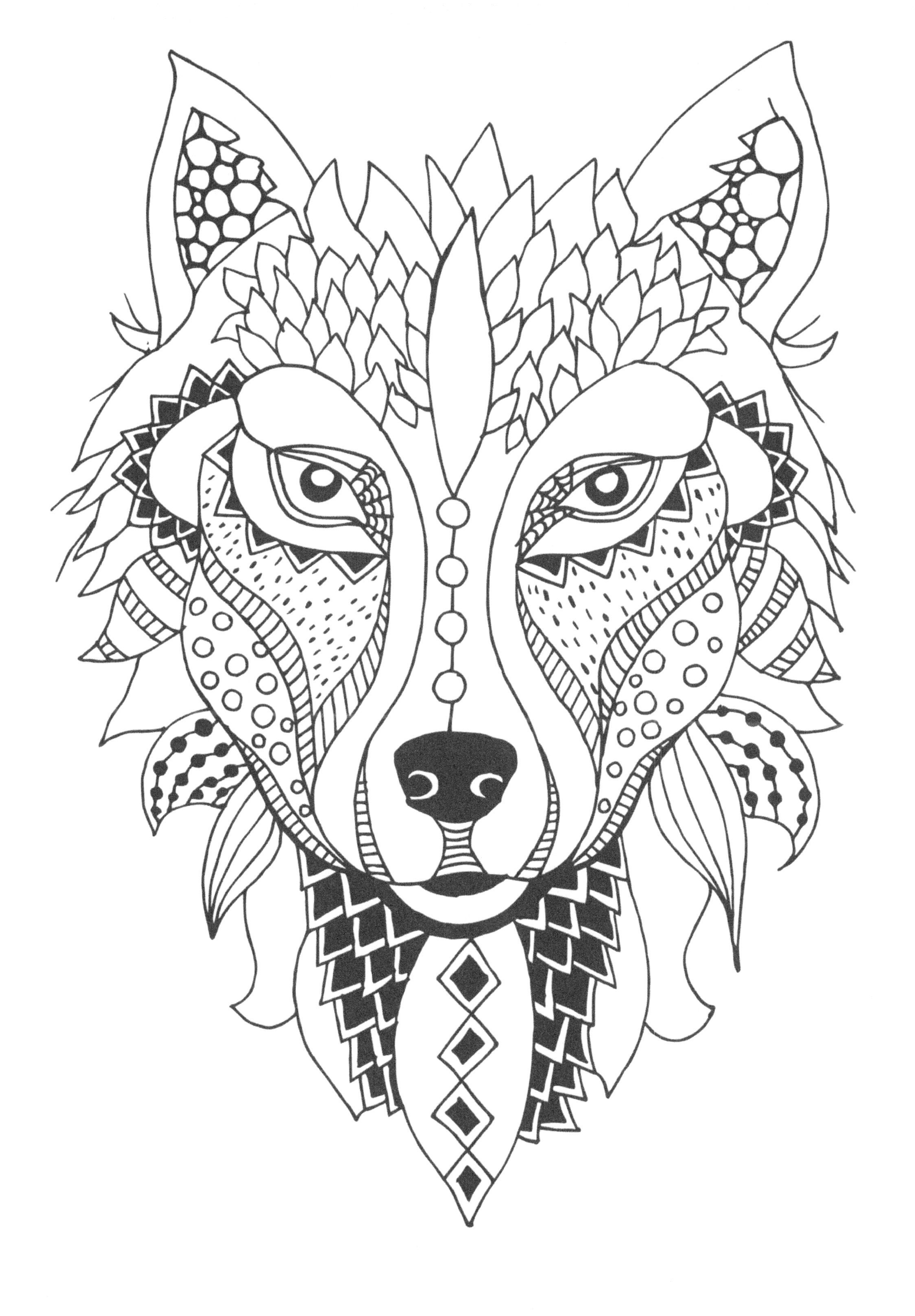

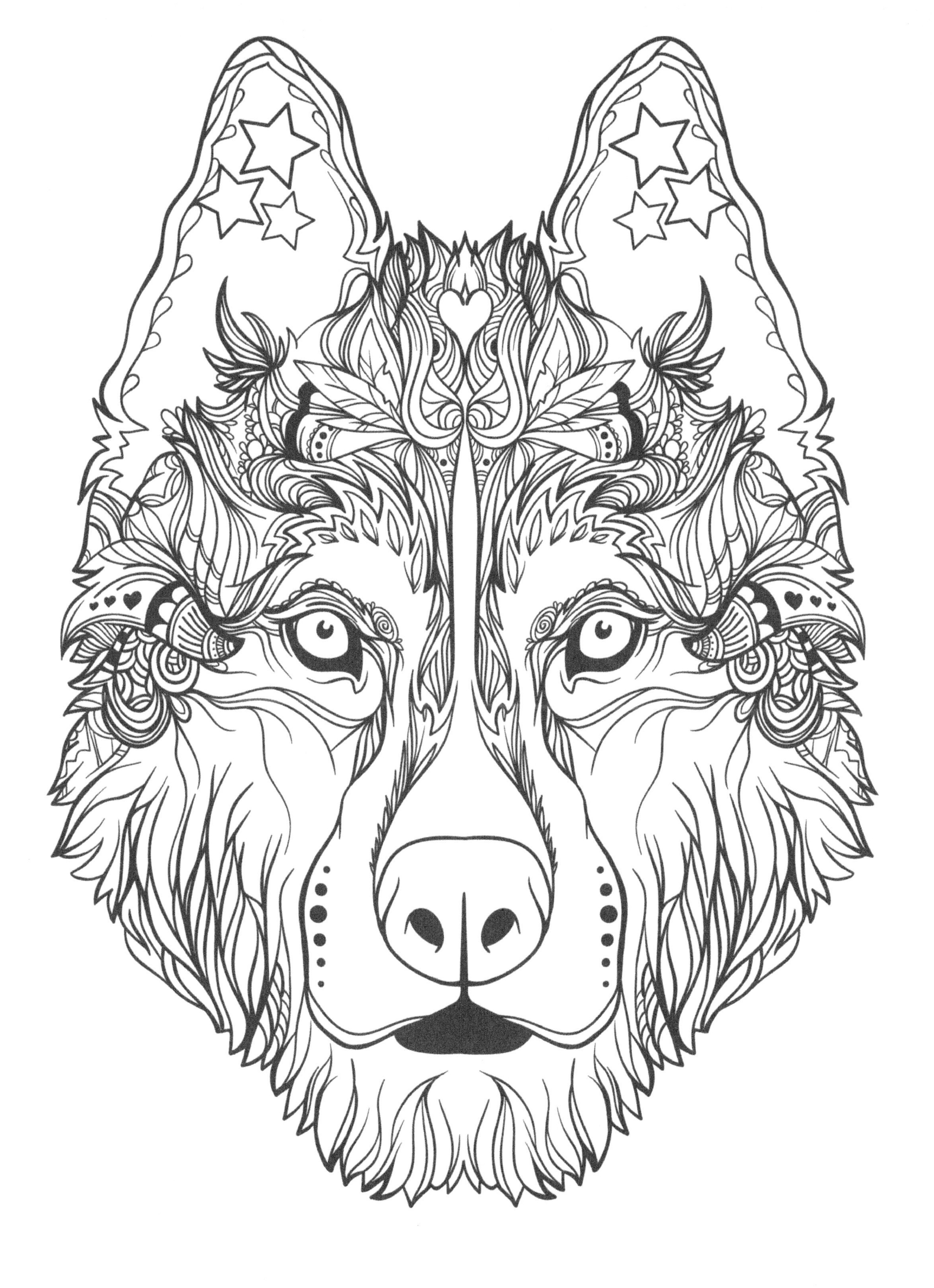

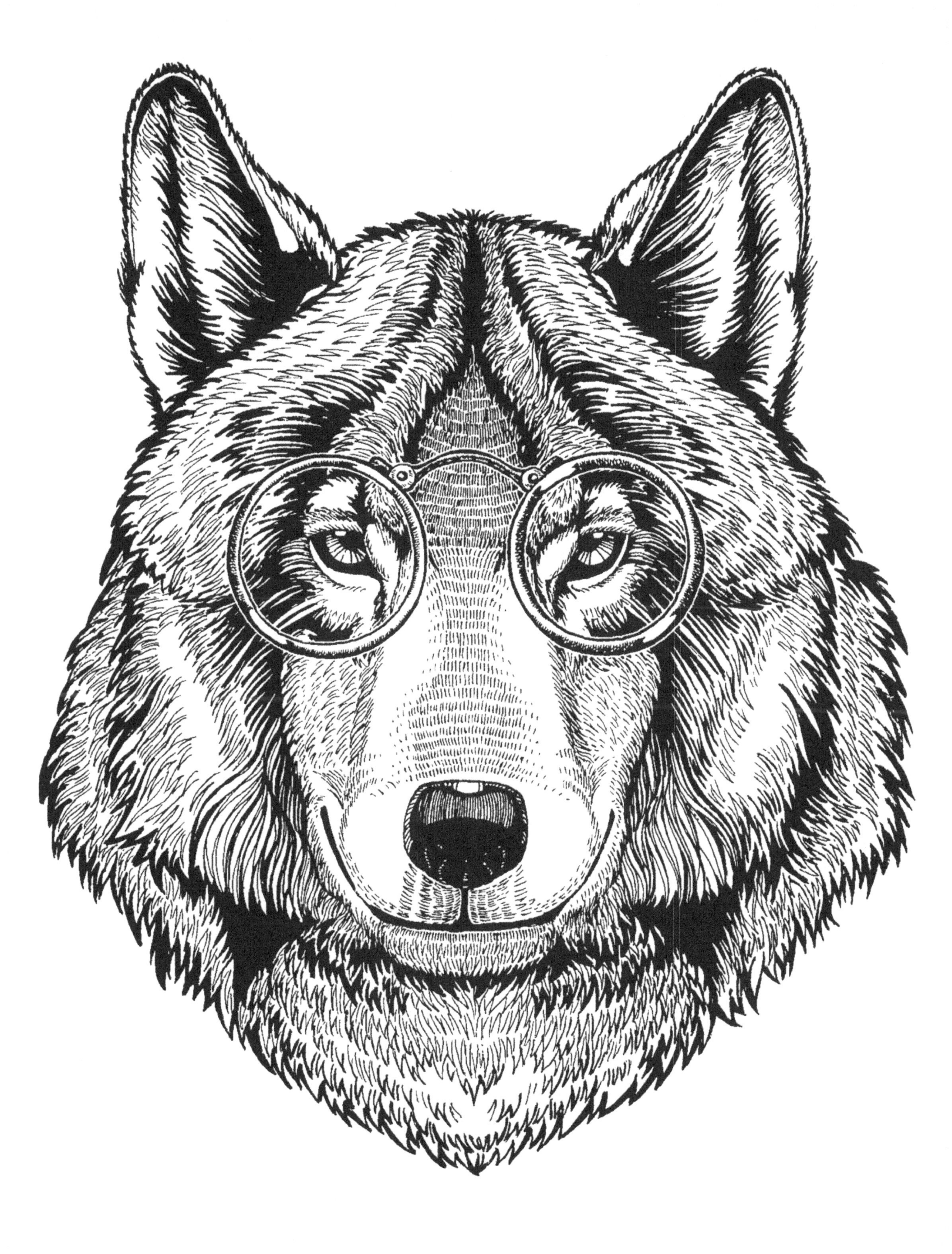

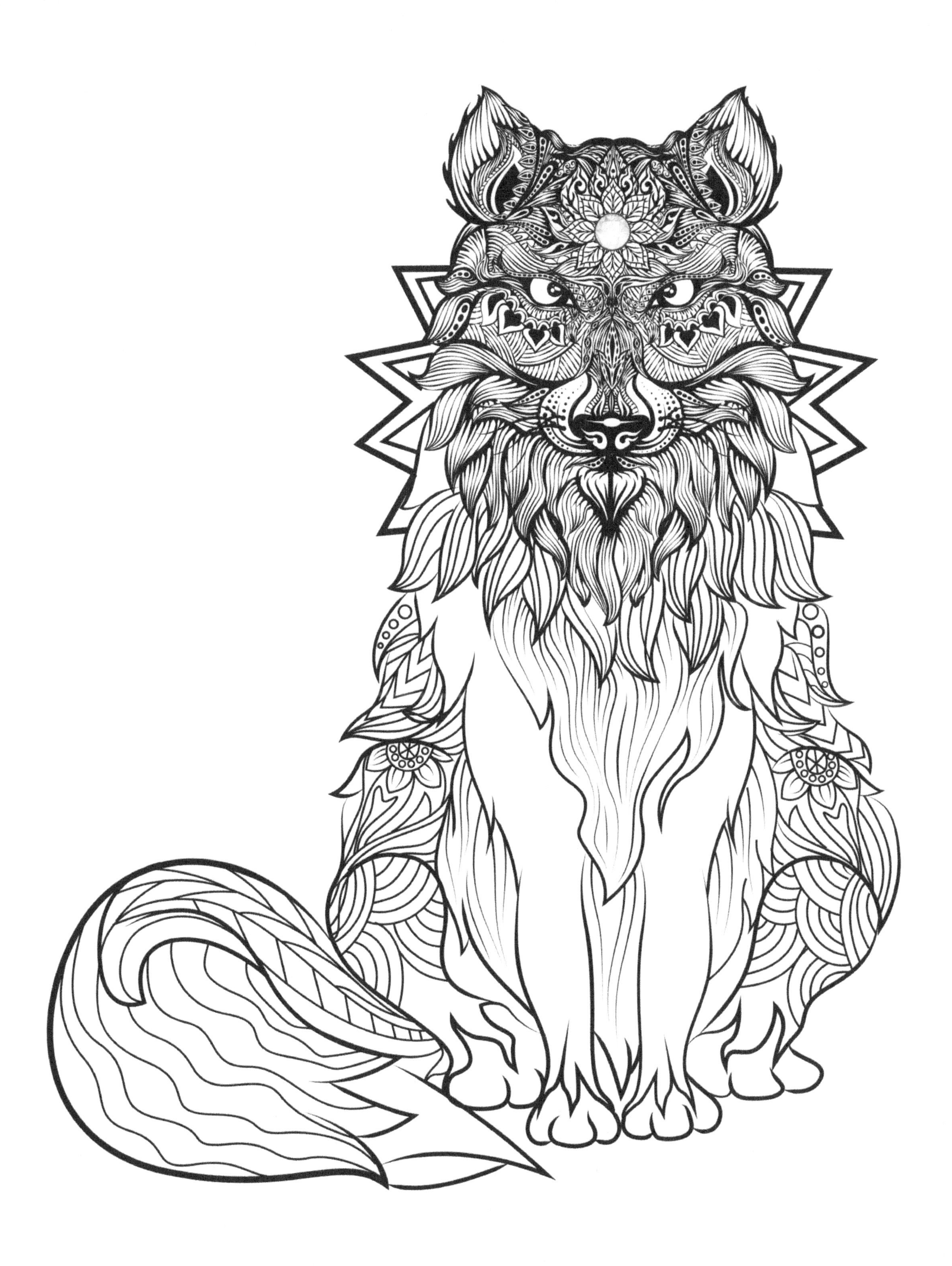

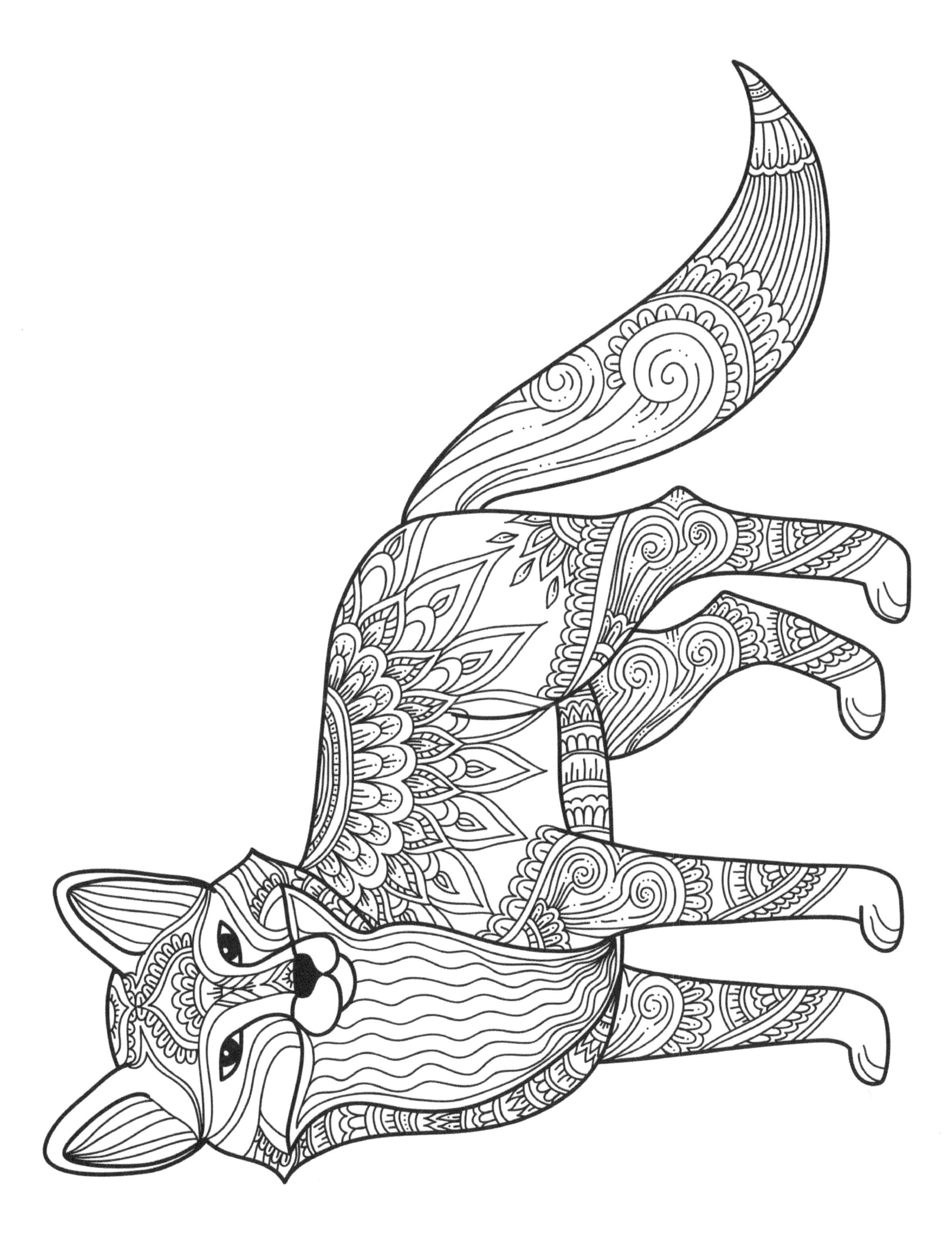

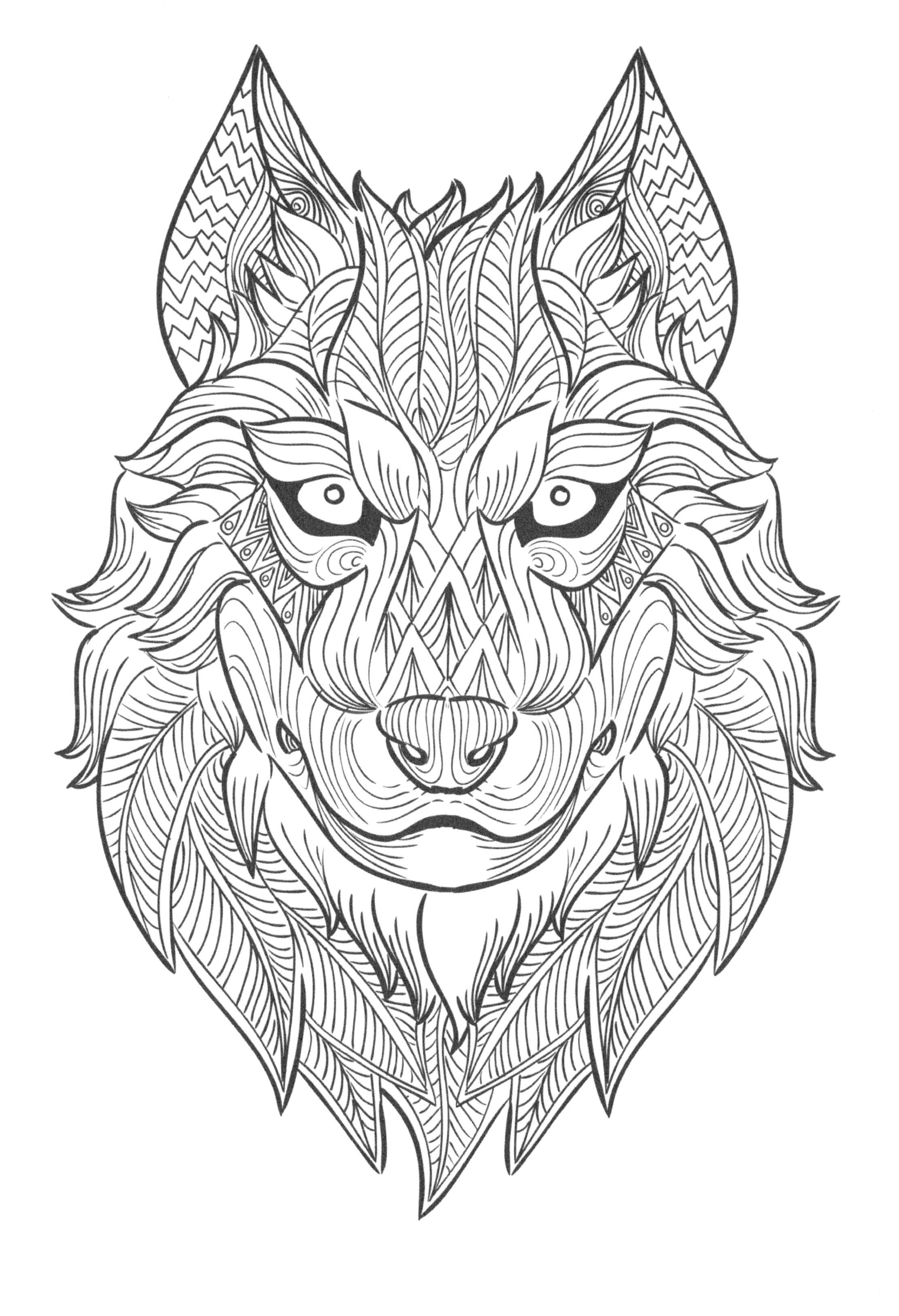

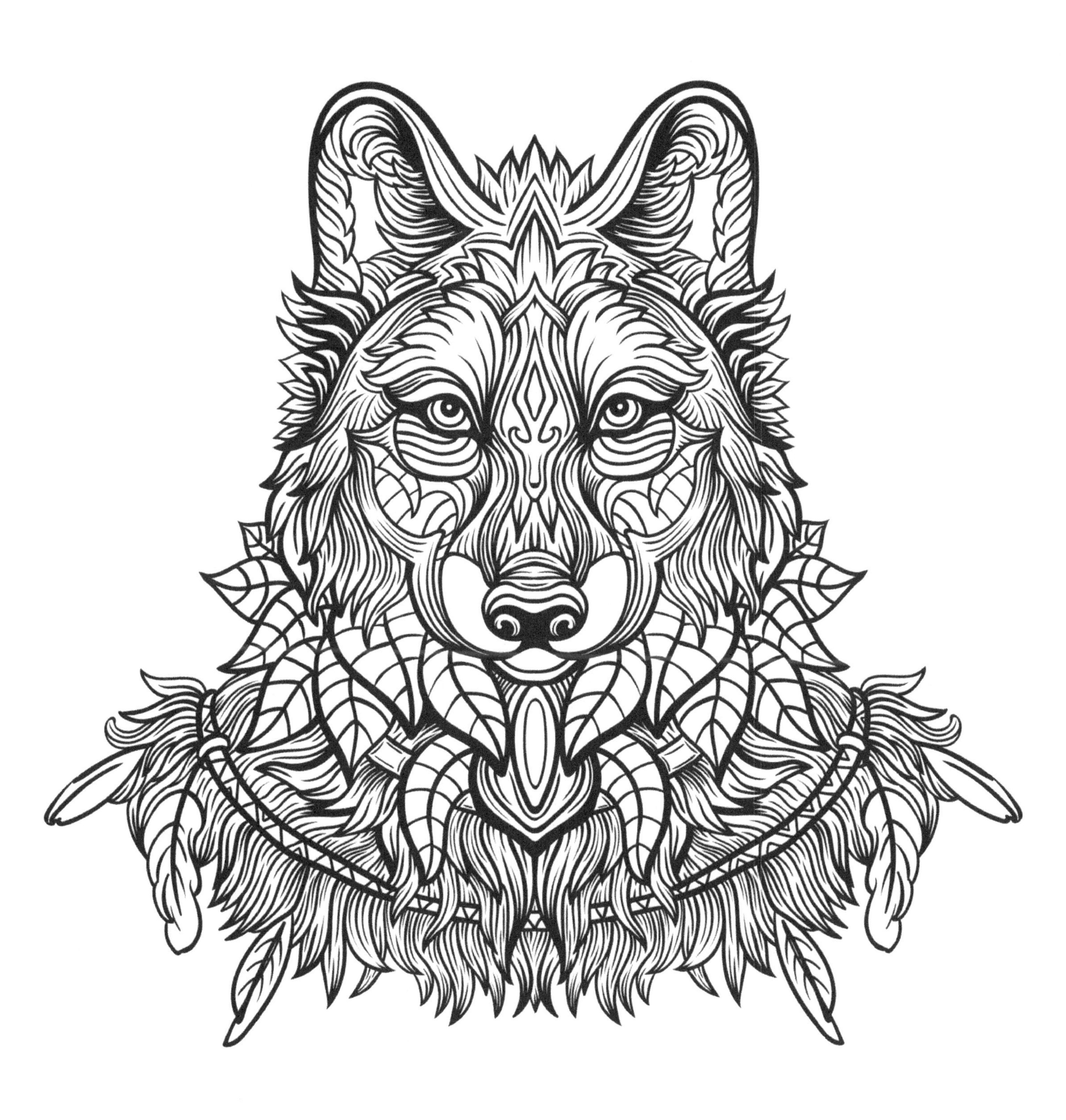

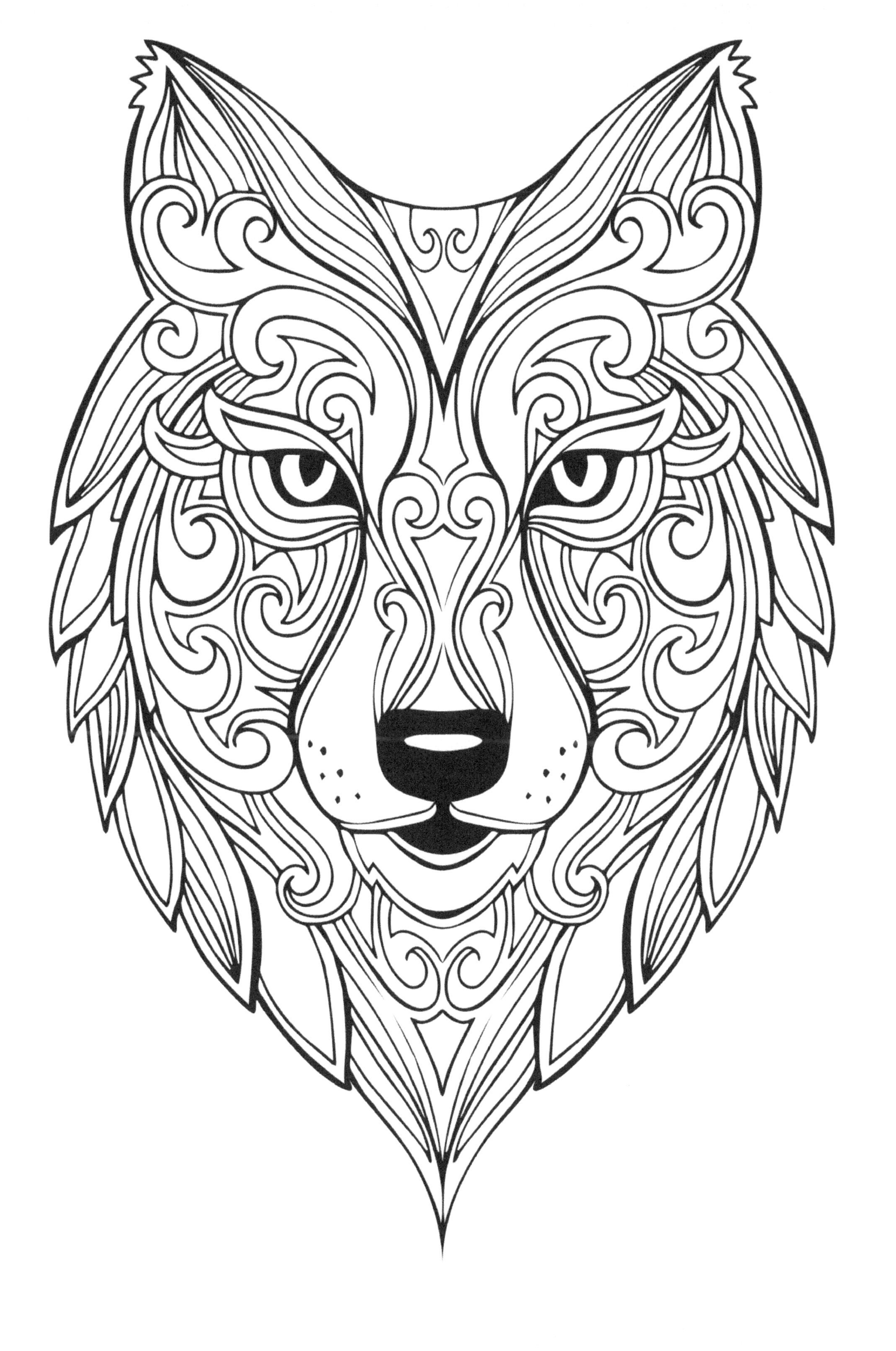

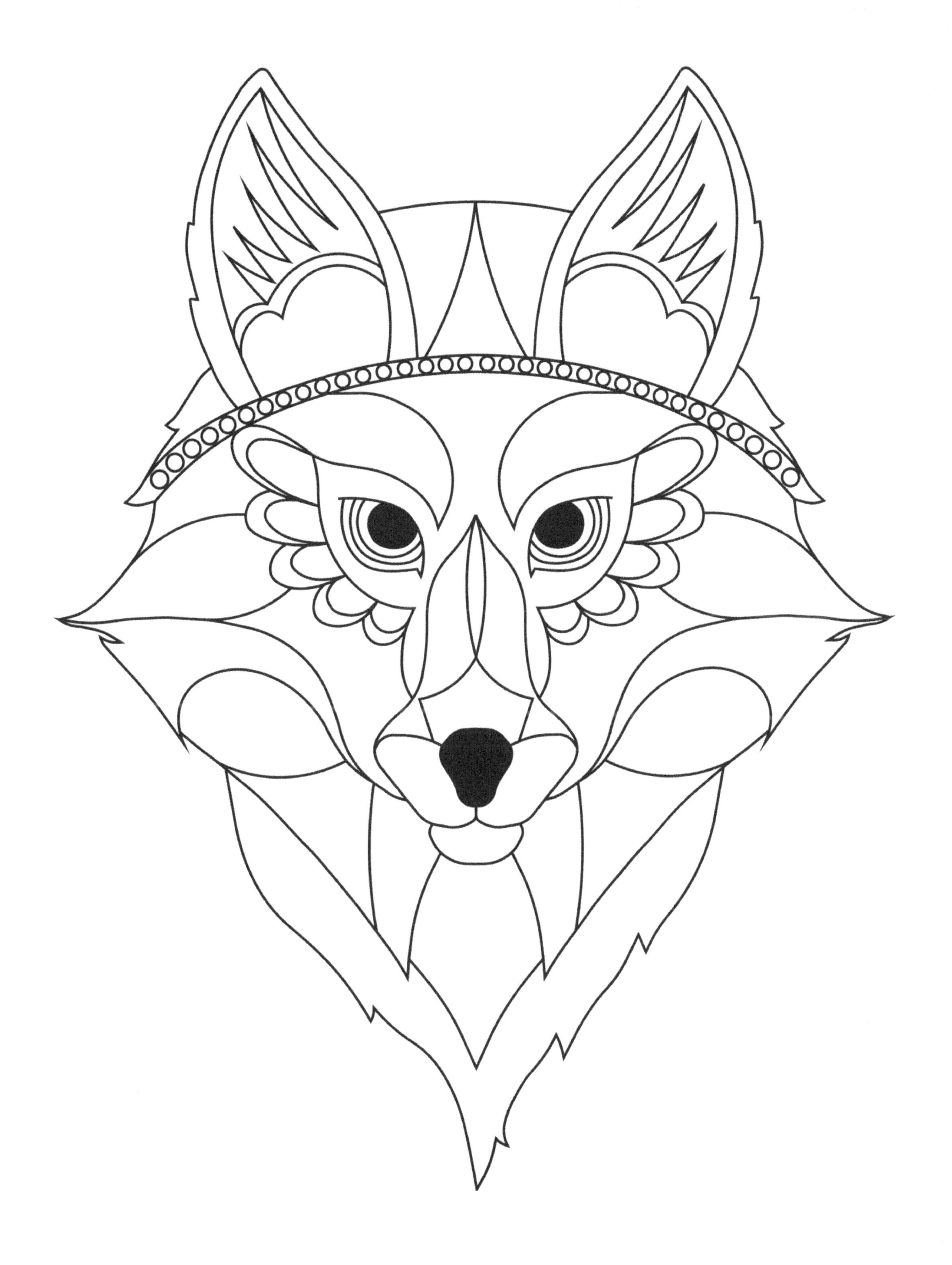

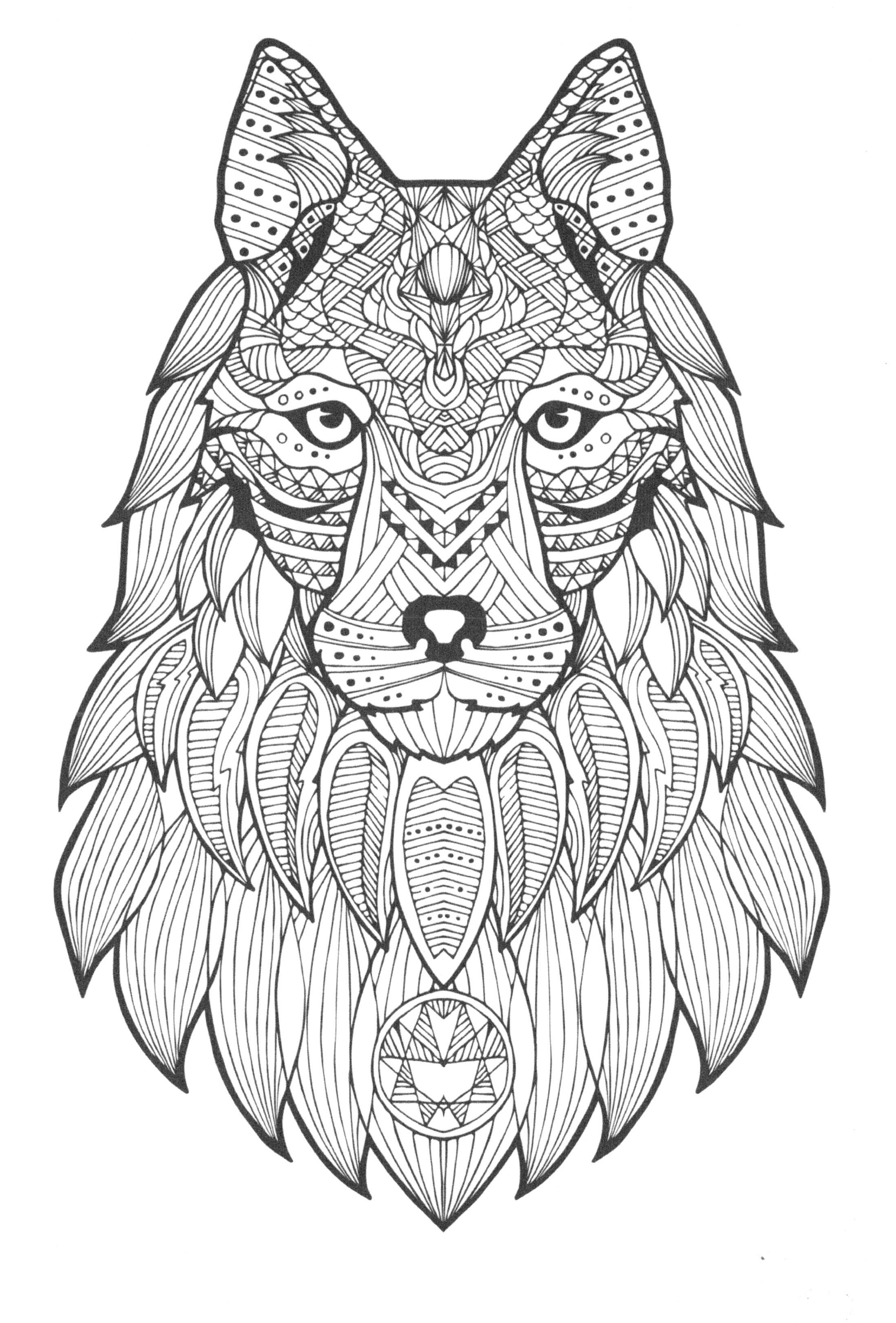

Made in the USA
Coppell, TX
12 February 2020